# Amazon Dot:

*Master Your Amazon Dot User Guide and Manual*

**Andrew McKinnon**

*Andrew Mckinnon*

## © Copyright 2016 - All rights reserved.

In no way is it legal to reproduce, duplicate, or transmit any part of this document by either electronic means or in printed format. Recording of this publication is strictly prohibited, and any storage of this document is not allowed unless with written permission from the publisher. All rights reserved.

The information provided herein is stated to be truthful and consistent, in that any liability, in terms of inattention or otherwise, by any usage or abuse of any policies, processes, or directions contained within is the solitary and utter responsibility of the recipient reader. Under no circumstances will any legal responsibility or blame be held against the publisher for any reparation, damages, or monetary loss due to the information herein, either directly or indirectly.

Respective authors own all copyrights not held by the publisher.

<u>Legal Notice:</u>
This book is copyright protected. This is only for personal use. You cannot amend, distribute, sell, use, quote or

*Andrew Mckinnon*

paraphrase any part or the content within this book without the consent of the author or copyright owner. Legal action will be pursued if this is breached.

Disclaimer Notice:

Please note the information contained within this document is for educational and entertainment purposes only. Every attempt has been made to provide accurate, up to date and reliable complete information. No warranties of any kind are expressed or implied. Readers acknowledge that the author is not engaging in the rendering of legal, financial, medical or professional advice.

By reading this document, the reader agrees that under no circumstances are we responsible for any losses, direct or indirect, which are incurred as a result of the use of information contained within this document, including, but not limited to, —errors, omissions, or inaccuracies

•

# Table of Contents

Introduction ................................................................... 7

Chapter 1 Revealing Your Amazon Dot ............................. 11

Chapter 2 Setup ............................................................. 13

Chapter 3 Hardware Basics and Other Settings ................ 21

Chapter 4 Using Your Amazon Dot ................................... 31

Chapter 5 Connecting to a Bluetooth Speaker ................... 65

Chapter 6 Troubleshooting ............................................... 69

Conclusion ..................................................................... 75

*Andrew Mckinnon*

# Introduction

I want to thank you and congratulate you for downloading the book, *Amazon Dot*. This book contains all of the latest production information on the small but smart revolutionary product known as the Amazon Dot.

Being able to choose the right speaker can turn into a rather complicated and overwhelming task. There are many different kinds of speakers and many different types of design to choose from. Some are more powerful and expensive while others are more inexpensive but don't provide the quality of sound you might be looking for.

Overall, every speaker serves the same purpose: allowing people to enjoy the music they love. Speakers can be more than just ordinary speakers: They offer the ability to have small voice-activated commands and various other functions. With many different functions to choose from, making the right decision about which product to purchase becomes even more difficult than before.

One of the most popular speakers on the market right now is Amazon's Echo Dot. The Amazon Dot is one of the newest hardware products available from Amazon. It is similar to the other products already produced by Amazon: the Echo and the Amazon Tap. While all three of these are all great devices to have, the Amazon Dot stands out in size and the ability to have voice-activated commands. Because of this innovation, the Amazon Dot is considered to be a smart gadget since it has the all the intelligence and functions of the Echo, but it is much smaller in size. This alone makes it easier to place in any room of the house.

The question, then, is: What makes the Amazon Dot better than the Echo? The Amazon Dot allows you to connect to all the other speakers in your living space, giving you a greater sound experience and a wide range of volume rather than relying on the device by itself. You can then control the type of sound around you by connecting it to an existing speaker system. This ensures that you can get the most out of this small device. The physical appearance and stunning design of this speaker makes it look like a small part of the Echo that has been detached from the big speaker. This allows a person to choose between buying a control system with the speakers or buying only the control system and being able to attach it to another device for more sound quality. Both of these devices are very good, but the Amazon Dot is more efficient with space and it had more abilities than the Echo.

By reading this book, you will learn about all the great features that you can expect from the Amazon Dot. From opening the box to troubleshooting, each page of this book will provide a thorough explanation as to how this device will work best for you. This will allow you to spend more time listening to music and using your Amazon Dot rather than reading through hundreds of other troubleshooting pages for this device. After reading through all of the product's pros and cons, it will be even easier to see why purchasing the Amazon Dot is a better choice than buying another kind of speaker system for you and your family to enjoy.

Thank you once again for downloading this book, and I hope that you will understand everything and appreciate all the things that are written in it. Enjoy!

*Andrew Mckinnon*

# Chapter 1

# Revealing Your Amazon Dot

Now that you have purchases your new Amazon Dot, the first thing to do is check the box to make sure that everything is inside. Please make sure that the box has all the below-mentioned content, otherwise contact the seller immediately. Since it is a useful device, it is best to get started with it right away with all the necessary pieces. Nothing is more frustrating than realizing you are missing a component necessary to set up your new device, so make sure that you have everything that you need.

**Inside Box**

Opening the Amazon Dot box, you should find following things inside:

- Amazon Echo Dot

- Micro USB Cable
- Quick Start Guide
- Power Adapter
- 3.5mm Audio Cable Output

**Uncovering Amazon Dot**

Like other Amazon products, Amazon Dot is covered in a protective foil to keep it safe from shipping and handling damage. Please take extra care while peeling off the protective gear. Do not use any sharp object like a knife or blade to peel it. By using care, your Amazon Dot will remain safe from scratches or any other damage that you may cause.

# Chapter 2
# Setup

Setting up the Amazon Dot is relatively easy as compared to other devices. Below are the easy-to-follow instructions and guidelines to facilitate your getting started with your new Amazon Dot. Please read carefully so you can set it up quickly and begin to enjoy all the features that it has to offer.

## Alexa App

Before anything else, you need to download the Alexa app to the device that you are planning to connect with your Amazon Dot. The Alexa app is free, so you do not have to worry about paying.

The Alexa app is compatible with various web browsers. Whether you are going to use your laptop, desktop, smartphone, or another handheld device to connect with the speaker, you can download it online. Please make sure to check the version of Alexa app that is compatible with your device before downloading. You can download Alexa from the Amazon App Store for Amazon-powered devices, Google Play for Android devices, and the Apple App Store for iOS devices. It is important to know that the Alexa app is not supported on the first- or second-generation Kindle Fire.

The Alexa app works with Amazon-powered devices that use Fire OS 2.0 or higher, Android-powered devices that are Android 4.0 or greater, and iOS devices using iOS 7.0 or higher. If the device you are going to use is lower than those described above, you may have to update it first.

**Turning It On**

It is recommended that you put the Amazon Dot in the location where you think it will be used most frequently. Please make sure not to put it anywhere near doors and windows so that it will not be hit and get damaged. Plug the power adapter into the Amazon Dot. You will know that it has turned on when the light ring of the Amazon Dot becomes blue in color. It will then turn orange and Alexa will greet you.

To be able to properly setup the Dot, connect the power cord of the device to the bottom and put it down where you might use it the most. Remember that it is best to find a place that is at least ten inches away from any walls, windows, or any other kind of obstructions. Once you have found the right place for the Dot to stay, plug it into the wall. As stated, a LED light ring will appear on the top of the Dot. It will go from blue to orange and then the Dot will greet you for the first time through Alexa.

**Using the Buttons on the Dot**

On the top of the cylinder, there are two buttons. The first button is the action button and the second is the microphone button. For the microphone button, simply press it down once to mute the device. You will know it worked because the light will turn red. Press it down once more to turn the microphone back on. With the action button, you will be able to wake up your device, turn an alarm on or off, and set up the Wi-Fi.

At the bottom of the cylinder is a LED light bulb. It is just above the wire that is connected to the power socket. When the device is completely connected to Wi-Fi, the light will turn white. If it is not connected to Wi-Fi, then the light will turn orange and you will not be able to access Alexa.

## Connecting to the Internet

You can connect your Amazon Dot to Alexa with the use of your gadget. Some instructions are there that are easy to follow. There are few things that you have to remember:

- If your Amazon Dot does not connect automatically, you can press the action button on your Amazon Dot for about 5 seconds.
- Open the Alexa app and press "Set up a new device" on the settings button.
- Your Amazon Dot will be connected to the Internet at this time.

It is important to connect only to dual band Wi-Fi networks. Do not use any kind of mobile hotspots for the Internet connection on the device. It will also not work if the network connection is set up as a peer-to-peer connection. To update or connect to the internet for the first time, follow the same directions as above. When you press the action button on the device, the light will turn orange and a list of all the possible network connections will appear. Select your Wi-Fi network and type in the password if required. If your network is not visible to other people or not visible on the list, scroll down to the bottom of the list and press "Add a Network." Type in your network and follow the same steps as above. If this does not work, then simply re-scan until the network you are trying to find appears. To finish everything, click on the "Connect" button and Alexa will be ready to use.

## Understanding and Using the LED Light Ring

As you will see, different colors on the LED ring say what the Amazon Dot is doing at that moment.

- If there is a revolving cyan light above a blue kind of background, the Dot is starting up.
- If there is a cyan light with a blue background behind it with the color in the direction of the person speaking, the Dot is processing the request and is busy.
- If there is a clockwise rotating orange light, the Dot is attempting to connect to the Wi-Fi.
- If there is a red light, the Dot is on mute.
- If there is a white light, the Dot is working to adjust the volume.

- If there is an oscillating purple light, the Dot is attempting to detect any errors while setting up the Wi-Fi connection.
- If there is no light, the Dot is waiting for your request.

**Speaking to Alexa for the First Time**

Now that you have set up your Amazon Dot, you will be pleased to know that you can start talking to Alexa. The first thing to do is to set up a wake word. Say the wake word "Alexa," then speak your command to Alexa afterward. Say things such as "Alexa, play me some music" or "Alexa, what is the weather like today?" Once your voice is fully registered and has completely reached the Dot, the circular blue light will appear.

Alexa will continue to listen throughout the day, will analyze the command, and will reply back to you. She will listen and wait to hear the wake word through one of the seven different microphones. This design was made for ease and comfort, so the user will not have the need to raise their voice. The wake word can be heard over other conversations or other music playing. If many people are trying to ask her a question, she will be able to pick out the different accents and voices of the people. She will separate all the questions and she will respond with phrases such as "music playing" and "the weather is sunny with day temperatures of 70 degrees."

If you want to change the wake word to personalize the device, go to the nearest phone or computer to open the Alexa app. Go to the control panel and you will see a list on the left. Select settings on the specific Dot device you want to change the wake word for. The wake word will be listed for each device. Select the wake word you want to change and create your new wake word. After you have done so, click "Save." Remember that this is only possible with the Echo and Dot from Amazon. As you have chosen your wake word, keep using it every time you want to give a command. If you ever need to change this, go to settings.

## Connecting to Another Speaker

Although Amazon Dot comes with a built-in speaker, you can still connect it to an external speaker if you are looking for better quality. You have a choice to connect your Amazon Dot to any speaker that you are going to use, but it is not necessary.

## Placing Amazon Dot

Now that you have learned all the basics of setting up Amazon Dot, you can put it anywhere you want. The sleek and compact design of the Amazon Tap allows it to be easily placed at various locations. Since it already has its very own built-in speaker, you will have a pleasant experience of hearing high-quality sound even without connecting it to any external speaker; it will allow you to listen to music anywhere at your ease.

# Chapter 3
# Hardware Basics and Other Settings

You need to know more about how the Amazon Dot works in order to have a better understanding of its functions. You may be surprised that this unusually small speaker packs a lot of features that make it much more than just a little speaker. It makes a great little device to have around the house since it is easy to use to create the atmosphere you want. You can listen to the kind of music you want to hear and you can ask it questions as if it is your own mobile device or computer. It simply takes out a few steps in the process and allows for the user to get to their music selection in one simple command. After a single request, the house can feel less empty, you can relax, or you can feel ready to work on the next assignment for your job. It creates an atmosphere that is unique for you.

Below are some basic hardware features that you may need to use frequently while using Amazon Dot.

**Working with the Remote**

The remote that comes with the device has a built-in microphone, a talk button, and other playback controls. The playback button can play, pause, increase volume, decrease volume, switch to the next song, and switch to the previous song. To talk to the device through the remote, press the talk button and talk into the top of the remote. The remote has its own kind of advantages. It is then able to wake up without a specific personalized word. Instead, you use the word "Alexa," or you can change it to "Amazon" or "Dot." For users who have purchased more than one Dot, the difficulty of responding to a certain device comes into play. If a person wants to speak only to the one of the living room instead of the other one in the kitchen, it becomes harder with the remote since, the remote will try to connect to all the devices in the house. To avoid this confusion, it is best to use different wake words or different start-up words for each device. This way, each device can have its own personal signal to attach to. Turning on the remote might send a signal to each of the devices in the house, but using a specific word will make it possible to speak to only one of the devices.

## Purchases for and within Your Device

One of the unique features of this product is that it allows you to buy both digital and physical products directly from Amazon using the Alexa features. This still uses the same one-click method as shopping on Amazon and you still need to provide a proper billing address, a payment method, and a Prime membership. With all of these, you will be able to activate the voice transactions and purchase products through your device.

Once you have successfully registered the Alexa device through the Dot, remember that the Voice Purchasing option will then be on by default. You can use different kinds of voice commands to carry out some of the common actions that are followed through ordering on Amazon, from purchasing another Prime eligible item, to reordering an item, to adding an item to the cart, to tracking the status of a previously ordered and shipped item, to canceling an order.

To enable or disable the voice purchasing options, activate a four-digit confirmation code, check your payment method, or check your billing address, you have to take the following steps. First, open up the Alexa App. Tap the settings button and click on the voice purchasing option. From here, you can access all the different purchase settings to make any change required. At the moment, there are a few categories of products that are still not available to purchase through the voice purchasing option. These categories are apparel, Prime pantry, shoes, watches, Prime now, jewelry, Amazon fresh, and add-on items.

With the voice purchasing options completely set up, it is time to go over the commands that can be used. This will allow you to make your own purchase through the Amazon Dot. The commands for this are "Alexa, order (item name)," "Alexa, reorder (item name)," "Alexa, add (item name) to my cart," "Alexa, track my order," "Alexa, where is my stuff," and "Alexa, cancel my order."

With the new way of ordering comes a new way to organize and view your cart to see the items waiting to be ordered. To manage your shopping list on the Amazon Dot, simply tap on the main menu in the Alexa App, select shopping and to-do lists, and you are on your way. From here, you will be able to add, remove, and edit the items on the list inside the app itself. You also have the option of doing all of this through the voice command, exporting the list to Evernote, Gmail, Todolist, or even iOS Reminders. This is all done by using the IFTTT recipes that is connected through the app.

## Microphone Off Button

To turn off the microphone, you need to press it for a few seconds. You will know that it has been turned off when the button turns red. Activate the microphone again by pressing the same button.

## Light Ring

The light ring indicates the speaker's current status; you can rely on the light ring to tell you whether Alexa is listening to you or not. Each light color represents a different state. These colors are explained later in this chapter.

## Volume Ring

Increase or decrease the volume with this volume ring. To increase the volume, turn the volume ring clockwise; to lower, turn the volume ring counter-clockwise. The volume level will be displayed in white to tell you the current volume setting.

**Action Button**

There are many different functions that the action button can perform. For instance, it can turn off the timer or the alarm that you have set beforehand; it can also wake your device and enable Wi-Fi mode. To activate the Wi-Fi mode, you need to press the button for five seconds.

**Power LED**

If you would like to know the current Wi-Fi status of the Amazon Dot, the power LED will give you all the details. Solid white light shows that the device has Wi-Fi connection. Solid orange light indicates that it is not connected to the Internet. Blinking orange light means that there is a Wi-Fi connection, but the device cannot connect to Alexa Voice Service.

**Micro USB Port**

The micro USB port gives power similar to what phone chargers can provide.

## AUX Audio Output (3.5mm)

This allows the device to be connected to other speakers.

## More about the Light Ring

It is important to know more facts and details about the light ring so that you will know what the device is telling you.

- **Continuous Violet Light:** Indicates that there is an issue with the Wi-Fi set up in general. Probably device is not connected to a Wi-Fi.
- **Solid Blue Lights:** The device is starting up.
- **No Lights:** This means that the device is waiting for your request.
- **Solid Blue with Light Directed toward Voice:** Processing the command that you have given.
- **Clockwise Spinning Orange Light:** Connecting to chosen Wi-Fi network.
- **Solid Red Light:** Microphone is turned off.
- **White Light:** Adjusting the volume level of your device.

## Changing the Wake Word

The default wake word that you can use is "Alexa." Changing it is easy. There are various reasons why you may want to modify the wake word. You may do it to fit your current needs, or you may also change it so that it will not be too obvious and remains unique.

If you have other Amazon Dot's in your household, stating just one wake word might activate all of them. You might want to personalize each Dot's wake word.

Go to the settings menu of the Alexa application and choose the device with the wake command that you would like to change. Scroll down until you see the space that says "Wake Word" and click on it. Do remember that, when you are changing the wake word, you just cannot change it to any word that you like. There is a drop-down menu that will allow you to make a choice.

It will take a few moments for the wake word to change and you will be unable to use your device during that time. Once you have changed your wake word, please write it down somewhere in case you forget it.

## Managing Your Lists

Using the Alexa app makes it easier for you to manage the lists that you have created. You can place various tasks on your list and you can also make a list of the items that you would like to purchase.

You cannot create multiple lists for each household member. Only one list can be created. Up to 100 things that can be included on the list. Listed items are limited to 256 characters. With this said, it is important that you keep your commands short and sweet.

If you would like to manage your list, you may do the following:

- Open it through the Alexa application.
- Check Amazon.
- Administer the list through voice command.

If you are going to control it by your voice command, you need to use commands that are easy to understand. You may say, "Add (item) to shopping list" or "Add (task) to my list." If you would like to know what you have written so far, you can ask, "What's on my list?" Alexa will give you accurate information every time.

If you would like to manage the lists that you have created through the Alexa Application or Amazon, here are the steps that you have to take:

- Open the Alexa app or Amazon.
- Select Shopping List or To Do List.
- View the list.

You may also add an item to a list that you have already created. Just select the + icon and add the new item that you are going to shop for or the new task that you have to do. By using the Alexa app, you can also mark things as they are done and you can delete tasks that you have already completed.

# Chapter 4

# Using Your Amazon Dot

The best thing about the Amazon Dot is that it gets smarter every time you use it. "Being smarter" means that every time you use Amazon Dot it will be more adjusted to your lifestyle patterns. Amazon Dot will recognize your voice more easily, and it will be more familiar with the commands you usually use. Ultimately, you will start considering Amazon Dot as your best companion because it will adapt your voice commands. Let's have a look at few of the bright features of Amazon Dot.

Many people who have purchased the Amazon Dot have discovered all its different effective features. This device can connect the Dot to your own home to make your house into more of a smart house connected to a command center. This command center is of course the Amazon Dot or the Amazon Echo, and it allows the person to control different parts of their home through the use of a simple voice command. This includes controlling the thermostat, Hue lights, garage door, and every other kind of Wi-Fi-enabled device that is connected to the Dot.

Now people can control their smart homes from anywhere by simply using a smartphone or tablet. All of the excitement begins in the morning after the fresh cup of morning coffee and setting up the lights in your home. The first feature to look over is how the device can lock and unlock doors using the Smart Dana lock recipe software. This kind of software on the device gives the user control over all the doors in the house, allowing the automatic locking or unlocking of any door at any specific time. All of this can be done through the use of the Amazon Echo, the Amazon Dot, or the Amazon Tap. They can all also connect to another kind of smart plug with a D-link. This connection through the plug allows for the device to connect to other devices such as the Netatmo Thermostat, Weather, Netatmo Welcome, and the On Hub. The refrigerator stands apart from the list since it has its own features. The GE appliance refrigerator channel allows a person to do many things with the fridge: For example, having the light blink on the fridge door if the door has been open for a long period of time. This can also be put in Sabbath mode, in which the fridge is open for a while but there is no noise attached to the device.

The coffeemaker can connect to the alarm and start brewing the morning beverage simply with a switch of a function. Now you can focus on getting yourself ready while your coffee prepares itself at the same time. The weather channel function allows the coffeemaker be set to start at sunrise. With this function, the machine can tell if the air conditioning in the house needs to come on simply through the detection of any humidity or temperature difference in the air. Knowing the device has these kinds of sensors brings more function and more efficiency to the household. When having guests over, the device can to greet them when they walk through the door. This way, you will not have to tell everyone about your new upgrade to the house. Instead, you can follow through will the original plans as the Amazon Dot will use Netatmo Welcome to change the environment when it senses that a new person has walked through the door. This way, even if you are gone and someone tries to get into your house, you can get a notification on your phone that someone is in your house.

But before you can set up this amount of security in your smart home, there are a number of devices that connect to Alexa that will use Alexa skills and IFTTT recipes to control the devices. The Amazon Dot will need to connect to more devices in order to have a greater range of coverage to see what is going on through the house.

## Basics of Alexa Commands

All of the various commands work with the Amazon Dot, Tap, and Dot. These basic commands are programmed to create an easy and more efficient way to use the device. They are meant to be short and easy to pronounce in order to give you more time listening to music and less time repeating yourself. These commands are: "Alexa, Stop," "Alexa, Volume (name a number zero to ten)," "Alexa, Unmute," "Alexa, Mute," Alexa, Repeat," "Alexa, Cancel," "Alexa, Louder," "Alexa, Volume Up," "Alexa, Volume Down," "Alexa, Turn Down," "Alexa, Turn Up," and "Alexa, Help."

With the last command, you can ask Alexa for help by asking her questions about features, music selection, and other forms of music that can be found. These questions come in the forms of commands as well. Simply say that word "Alexa" and then speak the question you want answered, such as "What can you do," "What are your new features," "What do you know," "Can you do math," "How can I play music," "How do I play music," "How can I add music," "How do I add music," "What is Prime Music," "What is Audible," "What is Connected Home," "What is Voice Cast," "How can I pair to Bluetooth," "How do I pair to Bluetooth," "How can I connect to my calendar," "How do I connect to my calendar," "What is an Alexa skill," "How can I use skills," "How do I use skills," "How can I set an alarm," and "How do I set an alarm." These phrases are broad enough to provide enough paths to the questions you might need answers to, and they are specific enough to be sure that the device can understand what you might be trying to say.

The next part to consider with the voice commands is a special feature called Alexa skill commands. Every Amazon Dot comes with a set of built-in abilities. These were created in order to enhance your experience with the device. You can customize and enhance these abilities on your own through the Alexa App itself. This will allow you to see new abilities for the Dot, add more abilities to the Alexa skill, and add more unique purpose to the device itself. All of these skills were developed by third party developers, and they are all the equivalent of the android apps on a typical smart phone. These skills are still in development, which means that not all of them will be available at this time. But, they are growing and developing fast through the industry as the popularity for the Amazon Dot continues to rise. Amazon is working to distribute this product to many other countries.

All of these skills are supported by a software term, "Launch." Once a skill is developed on the device, simply say "Alexa, launch (skill name)." Then the skill will be launched along with a welcome message for that particular skill. Some other information about the skill will be applicable with other sample commands included in the original message. To stop a skill is as straightforward as it sounds. Simply say "Alexa, stop," and the command will stop. Even when the Dot is speaking, this can still be done. The trick is to have a louder voice than the device itself. When saying the command, say it louder than the device so that Alexa can hear you and register the new command. For help with any of this, there is even a skill for that. Simply by saying "Help" to the device, Alexa will begin to read out a list of particular skills that are included in the Skill's Help

Fire section. With all of these abilities, the device quickly becomes personalized as you continue to see more of the features provided.

**To Make Things Stop or Reset**

Any device has times of trouble and any device has been created to reboot. Whether it is automatic or manual, resetting a device is sometimes mandatory to keep the system running at the highest level of efficiency. The reset button for the Amazon Dot is located at the bottom of the device, where the power cord plugs in. Use a small paperclip and press the button for about five seconds and the Dot will reset.

Another feature to keep in mind is how to be sure that Alexa has completely stopped listening. There may be times where you don't want everyone trying to control the device at the same time. To make sure that this doesn't happen, press the mute button at the top of the Dot. The LED ring will turn red once the mute is officially on. You can unmute the device by pressing the same button again. During this time, if you want the device to make any kind of software updates, you can force the update by keeping the device on mute for at least thirty minutes. Otherwise, all updates happen automatically.

**Everyday Life with the Dot**

Thanks to the Amazon Dot, there is no need to be constantly pulling out phones or using computers for information on the simplest things. Wanting to know your schedule, the weather, and common house information should only take a few seconds instead of a few minutes. Pulling out the phone and looking up what the weather is like for the day takes longer than necessary once you have the Amazon Dot. Most things should now require only the time it takes for you to ask a simple question.

One thing this device can do is basic math operations. Simply adding numbers or converting units is very helpful in the kitchen or when working on financial statements. Speaking up and asking out loud what a number is or what the answer to a certain equation is proves in itself to be a great tool to have in the house. For example, you can ask "Alexa, what is one thousand eight hundred seventy six divided by four," "Alexa, three point four eight six times twenty four," "Alexa, convert twelve feet to centimeters," "Alexa, convert seven tablespoons to teaspoons," "Alexa, convert thirty five degrees Fahrenheit to Celsius," and finally "Alexa, how many miles are in thirty kilometers." As you can see, these are some of the simplest things that some people cannot do in their heads. People normally do not memorize the conversion from a tablespoon to a teaspoon, but they might need it when dividing a recipe. People normally don't memorize multiplication tables involving three places for decimals, but they might need to know the answer when looking at budgeting or measuring out materials for a project.

For cooking conversions, it is the same as asking the other kinds of common questions above. For example, "Alexa, how many teaspoons are in two tablespoons," "Alexa, how many pints in a gallon," "Alexa, how many cups are in a single quart," "Alexa, how much is half of three fourths of a teaspoon," and finally "Alexa, how many tablespoons are in a quarter of a cup." While some of these may seem strange at first, Looking at recipes and working on different portion sizes for different people can cause major changes in the amount of ingredients. When changes are needed quickly, they only cause more confusion for people. Luckily, to they can rely on their smart devices for faster work and more efficiency.

To hear a flash update, people normally turn to the television or the nearest radio. You can configure the Amazon Dot through the app to bring you news from many different types of sources. It all depends on what you want to watch most and what you enjoy listening to most. Some news programs are longer and more focused on the economy rather than what is going on in your local town. It all comes down to preference since they all were created to run in different ways. The Amazon Dot can collect information and tell you additional resources from BBC, Economist, TMZ, and NPR. To hear these new stories, simply say "Alexa, what is my flash briefing." The device will then register what you said and it will play the news from the specific source you selected. If you want to configure the flash briefings, open the Alexa app and select the navigation panel on the left. Go to settings and select the "Flash Briefing" option. Then you will be able to customize the flash briefing by choosing different shows, news, headlines, and weather updates.

The Dot can tell you the scores from all the live matches and games for your own favorite teams, along with their schedules. Simply say "Alexa, what is the score of the (team name) game" or "Alexa, when does (team name) play." To get more localized information about a team, you can change this feature, too. But it only works in the United States. To do this, go to the settings in the Dot App and tap on the Dot device location. Enter the zip code of your area and tap "Save Changes." This feature will give personal weather reports, local news, and other pre-recorded shows that are seen in your own local area.

To get traffic information and find the fastest route has become more of a mandatory situation. Being able to set the fastest route on road trips and find the fastest way to work gives you more time and less stress on the road, creating a safer trip and an easy way to get to where you need to go. The Amazon Dot gives the most efficient and effective routes for a person to take. In order to get this to work, go to the settings of the Dot App. Tap on "Change Address" and input the current address in the "from" and "to" fields. Then remember to save the changes. All of this will give you the most accurate traffic information on the route you want to take. It will give the best route with the least amount of traffic to your destination.

This device can also read Kindle books through the speaker system. This feature allows you to immerse yourself in a book and relax while listening to it. You can read any book in your Kindle library simply by saying "Alexa, read my Kindle book," "Alexa, Read my book (title)," "Alexa, play the Kindle book (title)," or "Alexa, read (title)." As with the other commands shown in this book and instruction manual, all of the commands are simple to know and comprehend. This allows tasks to be done with more ease and less frustration of repeating yourself. With all the different phrases the device can register, it becomes easier to move through. This same process can also be applied to the audiobooks. A certain narrator reads through the chapters, so you are not continually listening to the voice found in the Amazon Dot.

**Knowing the Weather**

Everyone is curious to learn about the surrounding weather. It is an important way to plan the day-to-day activities so you know whether you need an umbrella or suntan lotion. The Amazon Dot will keep you updated with accurate weather parameters while you are in your bedroom, your kitchen, or wherever you choose to place your Amazon Dot.

## Use It as a Timer or Alarm

Amazon Dot comes with the unique feature of having an alarm or timer. This one feature makes Amazon Dot well ahead of its competitors. You will not find this feature in other similar devices. You can replace your alarm clock with Amazon Dot, as it is more convenient than traditional alarm clocks. You can set up the Amazon Dot as your very own timer and put it anywhere, depending to your requirements. Just set the time you want to wake up and Alexa will be more than happy to wake you at that time. You can also set a time for various functions to be performed by Alexa app. For example, Alexa will read all the latest headlines to you so that you know what is happening in the outside world before you leave the house, or you can set up a beautiful song for your loved ones at a particular time. Just imagine the convenience of not having to wait for the news to be delivered or being able to dedicate a romantic song to your partner at an appropriate time.

## Listening to Music

Please be aware that the Amazon Dot does not have the type of power that the Amazon Echo has because of the difference in size but, since it can be connected to any other speaker that you have, it can give you the ultimate music listening experience. At the same time, you do not need to worry about the songs that you hear. The Amazon Dot will provide you hands-free control for Amazon Music, Prime Music, Pandora, Spotify, TuneIn, and IHeartRadio. You can just ask the Amazon Dot to play the music that you want from the station of your choice. The best thing about Amazon Dot is that you do not have to press any button to do this. You just have to say your wake command and it will listen to you.

If you want to play from your device's playlist, make sure that the Amazon Dot is connected to your Alexa application. You can give commands to shuffle the songs or you can control the specific songs that are be played just by using your voice. If you want random music to be played, let Alexa pick out the songs to be played for you.

**Listening to Audio Books**

Amazon Dot will allow you to hear your favorite stories. If you miss those memories of good old times, you can get the same childhood feelings by listening to audio books. There are various books from which you can choose. You need to give the title of the story to Alexa and then you can simply let Alexa read the audio book to you.

## Have Control over Your Smart Home

If you have a smart home, the Amazon Dot fits right in with it. You can expect it to connect it to all the devices that make controlling your home much easier. Perhaps you do not want to get out of bed, but you would like to turn the thermostat up or down so you can feel more comfortable. If you want to control your television and you cannot find your remote, you can ask Alexa to change the channel for you. It will make living at home easier than before. The Amazon Dot can also control your lights, making life a lot easier.

To continue building your smart home through the Amazon Dot, it is best to connect it through the various devices found in your house. This can be done either through the Alexa App directly or through a smart hub. There are many common devices in the home that can be connected through the network. The first items are the lighting and fans. They will need to use LIFX Wi-Fi smart LED light bulbs and the Haiku Wi-Fi ceiling fans. With these, they can be controlled through the device and they work through the network connections in the house. For switches and outlets, you have to purchase the Belkin WeMo light switch, smart plug switch, and the insight switch. You will also need to purchase the TP link in the case of the smart plug and the smart plug with energy monitoring. The last thing to purchase for the switches and outlets would be a D-link Wi-Fi smart plug to give the largest amount of manual control of lighting and fans.

In this same category of controlling the smart house through the device, there are the Nest Learning Thermostat, the ecobee3 Smarter Wi-Fi thermostat, and the Sensi Wi-Fi Programmable Thermostat. For the locks in the house to be controlled through either the Amazon Dot or the Amazon Echo, there are the Garageio and the Danalock. The car control in the garage becomes completely automatic with these devices.

While this all may seem like a ton of technology to purchase and handle, remember that it is all optional. These optional methods are meant to provide a kind of ease for the consumer and to provide more of a connection between the device and your own routine. This same message has been repeated to prove the truth in the statement. Everything in life has upgrades and everything that is upgraded to the fullest extent is meant to give a wider vision to follow. It is not meant to give trouble, it is meant to help you turn your own house into a smart home.

## Connecting Smart Devices to the Amazon Dot

In order to fully connect these devices to the Amazon Echo or the Amazon Dot with Alexa, you must first create an account for the smart device. This is the common native app that is always relied upon and always used as a place to turn back to. The second step is to connect the smart device account to your own original Amazon account. Afterward, you will be able to use voice commands or the Alexa app in order to completely control the smart device.

As noted before, in order to have the complete ability to control devices through the Amazon Dot or the Amazon Echo, the common devices in the house will need to be replaced or upgraded. With the lighting in the household or your new smart home, the following items will need to be purchased: Philips Hue series, Philips Hue bridge kit, Philips Hue starter kit, Cree Connected LED Samsung SmartThings hub or Wink hub, GE Link bulb for the Wink hub, Osram Lightify smart bulb for the Wink hub, and the

TPC connected smart bulbs for the Samsung SmartThings hub or to be used again for the Wink hub.

For the outlets, dimmers, and switches in the house, there are more items to purchase. With these items, it will be easier to have a connection for the devices in the home to follow. The items to purchase in order to make this work are: iHome smart plug for the Wink hub, the Samsung SmartThings outlet for the Samsung SmartThings hub, the Insteon switches, the Insteon dimmers, the Insteon outlets for the hub, the GE Z-wave switches and outlets for the Samsung SmartThings hub, the Leviton switches, the Leviton dimmers, and the Leviton outlets for the Samsung SmartThings hub and for the alternate Wink hub, as stated before.

The same can be said about thermostats: More items have to be purchased to provide an upgrade for the smart home. The device will then be able to use a sensor to see when the air conditioning or the heater needs to be turned on and it will use the level of humidity to see what the temperature should be. In order for this to occur, the following need to be purchased: the Honeywell Lyric or Total Connect Comfort thermostats to connect to the Samsung

SmartThings hub, and the keen home smart vents through the hub.

## Smart Home Hub

With all the purchases and actions required, people might question why they would really need to have all of it. They might question the advantage of having an upgraded system. The main point is that these smart hubs are necessary for the connection between the house and the Amazon Echo or the Amazon Dot with Alexa. Many of the smart devices will lack the radio that is required for a direct network connection. Since the technology has not been created to have a kind of radio network transmission in each of the smart devices, a hub provides that connection.

With the connection completely set up for the device and Alexa to follow, you can see the amount of automatic control over the house. With the integration of a smart home hub network, there will be a link between the Amazon Echo or the Amazon Dot and another specific device you want it to connect to. With the efficiency of the device and the programmable links to be gained, there will be more ease and more control to gain in the household. The following is a list that shows the smart home hub devices that can be bought to control the smart home hub devices: Samsung SmartThings hub, Wink hub, Insteon hub, Philips Hue bridge kit, Philips Hue starter kit, Caseta wireless smart bridge, alarm.com hub, Vivint hub, Nexia home intelligence bridge, the Universal Devices ISY hubs, HomeSeer home controllers, Simple Control Simple Hub, and the Almond Smart Home Wi-Fi routers. While all of these devices have been listed before, this shows the overall plan of what to get and the basic idea of what is needed for the entire system to function.

In order to connect the hub to Alexa, it is first important to note that here we are showing specifically how to connect the Samsung SmartThings hub to Alexa. Any other kind of hub will have a similar procedure, but these instructions apply to that specific device to explain the process. Tap on the menu button in the Alexa app (it looks like three horizontal lines on the top left). In this menu, select "Smart Home" and scroll down to "Your Smart Home Skills." Select "Get More Smart Home Skills" and enter the word "SmartThings" in the search field. Select "Enable for SmartThings." Either register or log into the account using the SmartThings email and password combination. Once you have logged in, choose the SmartThings location from the menu provided on the device. Tap on the checkbox for every device that Amazon's Alexa will need access to. Select "Authorize." Once authorized, the following message will appear: "Alexa has been successfully linked with the SmartThings." Once you are done, tap on the small "X" in the corner to close the window. You can now begin discovering and seeing everything the device can do.

**Discover Devices**

With everything connected, there are multiple skills and many other discoveries to be made. Once again, open the Amazon Alexa app. After doing what you did previously and closing the success message, the app will guide you automatically to discover different devices. Select "Discover Devices" and then wait for the entire device discovery to be complete. If the automatic prompt does not come up, there is a second way to discover the devices. Open the Amazon Alexa app and tap on the menu icon. Select "Smart Home" and scroll down to your devices. Select "Discover Devices." Wait for the device discovery to be complete. Once the discovery has been fully completed, the discovered devices will be listed under the "Your Devices" section of the smart home. If the devices you are looking for are still not coming up on the list or you are still not getting the automatic prompt, see the chapter on troubleshooting or contact the sellers of the product.

To control these newly discovered devices, there are voice commands to be followed. They are as follows: "Alexa, turn off the bedroom light," "Alexa, turn on the bedroom light," "Alexa, brighten the kitchen light," "Alexa, dim the kitchen light," "Alexa, set the bedroom light scale to (brightness scale of zero to one hundred)," "Alexa, lower the kitchen thermostat by (number) degrees," "Alexa, raise the kitchen thermostat by (number) degrees," "Alexa, set kitchen thermostat to (number) degrees."

Now that the voice commands have been seen and the lists have been found to search for the discovered devices, you may still want to add more devices in the same process. After connecting everything through your Alexa app and you want to connect one more, you can still use the Samsung SmartThings app to update or change the devices from which the SmartThings devices Alexa will be able to control through either the Amazon Dot or the Amazon Echo. Whether you decide to add a new dimmer switch, a new outlet, a new on/off switch, or a thermostat to the existing SmartThings setup you can follow these steps to provide more control to your own home. In the SmartThings app, first select the menu button and tap on "Smart Apps." Select the Amazon Dot and you will see all the devices that Alexa can currently have access to or recognize. Tap on the "My Switches" or the "My Thermostats," then select all the checkboxes for each device that Alexa will need to have access to. Select done and then next. Tell the app "Alexa, discover new devices." Wait a few moments for Alexa to completely confirm and wait for the discovery to be complete. Tap "Done." Now the new devices should be added to the app and they should be fully functional to connect to the device.

Disconnecting SmartThings from the Amazon Alexa follows a similar process. Uninstalling Alexa from the SmartThings app will remove the connection between Alexa and the other smart devices. The smart devices will then not be able to respond to the voice commands in the SmartThings devices. To completely disconnect the features in the SmartThings app, start by tapping the menu icon. Then select the "Smart Apps" option. Select "Amazon Dot" and then tap "Uninstall." Afterwards, click "Remove" and confirm the removal. The next step is to disconnect more of the features within the Amazon Alexa app itself. Select the menu icon once again and then click on the "Smart Home." Scroll down to "Smart Home Skills." Select "Disable" for the SmartThings. Then select "Disable Skill" to confirm the final result. When all of this is completed, you have freed up space in the device and can focus on the devices that need to be connected.

**IFTTT Commands with Alexa**

One of the questions you might have asked yourself is: What is the IFTTT command that has been mentioned before? IFTTT is an interface that can provide the fastest and most efficient link to other various kinds of apps and functions that can follow the apps and functions with Alexa. It is basically the main software, or at least part of the main software, that can connect the device to other results. There are many different kinds of IFTTT recipes involved in the Amazon Dot and the Amazon Echo. These were created to automate the life of the user and carry out the different kinds of repeatable tasks that are required to be followed normally by the user.

All of this was created to have a network connection with each of the devices. They were designed for the purpose of saving time and effort. The first thing to do when setting up this kind of software is to connect the main Amazon account to the IFTTT. In order to do this, you again have to follow similar processes. The first step is to go to the IFTTT and start by setting up an account if you do not have one already. Once registered or logged in, go to the "Channels" home page and find the Amazon Alexa channel. After selecting this option, a prompt will appear on the screen and tell you to enter the Amazon account info and it will require you to sign into this account. Once you have signed into the system, you will be able to access all of the other kinds of existing recipes found by Alexa. You will be able to choose from over 800 options and add them to the account. With the IFTTT now completely set up, the next step is to utilize voice commands to automate the smart house.

**Using the IFTTT responses**

Before automating the house with this software, you must think about the phrase you want to use and the temperature at which you want to keep the house. The device will be able to sense the temperature and the level of humidity in the house. The software will be alerted to the change and will automatically change the temperature. To start this process, say the phrase "Alexa, set (phrase)." You can either say what you want a room temperature to be or where you want the Nest thermostat to be set.

Taking a step back, the first thing you need to do is connect to the correct Next channel for the IFTTT. Start by as always opening up the Amazon Alexa channel or app on any smartphone or computer. You can do this by clicking on the three horizontal lines in the top left corner of the device. After this has been done, scroll down and choose the "Smart Home" option. You should now be at the "Device Links" tab. Under this section, choose "Nest" and then continue. Log in or register to the Nest account with your own user id and password. You should now be able to see the discovered devices. If this Next is on the local points of the Wi-Fi network, Alexa should naturally be able to discover the device. After all this is done, you will be able to set the room temperature with the voice command. Simply say "Alexa, set the room temperature to (number) degrees." Now one of the main parts of the software has been completed and can be utilized.

## Using the SIGNUL Beacon Channel

Another rather unique way to create that bridge and the connection between the digital network and the physical devices is by using another kind of software that can be integrated into the device. It can connect the network to your own personal devices by detecting the presence or the absence of your own smartphone. You can define the zone entry and the other kinds of exit events for the SIGNUL beacon to follow. This will help it to maintain and automate other kinds of mundane tasks. In this channel, you are then able to use the physical context of the network streamline in order to connect to the digital world.

Once you have set up and created this channel for the device, the next thing is to open up a spreadsheet. This is a special kind of spreadsheet for the channel to follow and work towards. The slack on the sheet is then informed that you are ready for the software the moment you sit down and are ready for work at your desk. This same thing will happen when you come home, as the lights will turn on after it is alerted to your presence. Once you want to go to bed, it can see that you are asleep and it can mute your phone. When you decide to leave work, turn on the Next thermostat to repeat the process again. From this, you can complete these general items and they are then recorded on the spreadsheet. The SIGNUL beacon will then naturally pair to the IFTTT account after you have done it manually. Simply use your account to pair these two accounts together first.

With different levels of the house, there are many different areas in which to work and many different lighted areas. Sometimes the light on the top floor is left on when you move down to the bottom floor. After working in the attic, for example, you might want to get ready for bed and then you remember that you forgot to switch off the attic light. By grouping the lights of the house together, you will be able to turn off all the lights in the house through one simple voice command. This way, at the end of the day, you do not have to worry about lights being on and you can relax comfortably.

The Amazon Dot and the Amazon Echo both work by being linked to smart device, as seen in many of the previous cases before. Through the Wi-Fi network, Amazon can provide the most support possible to the lights and switches in your house. While this is only possible with a few of them, the rest of them can be controlled through the smart hub. The lights to be controlled by this hub are: the BR30 downlights, the Hue H19 traditional bulbs, the light and bloom strips, and the Lux white bulbs. If you are attempting to connect all the lights in the house together, you must look at both the lights and the switches to upgrade in the house. For the switches, the following must be purchased: WeMo light switch, Insight switch, and switch.

By using Wink to group together the lights and control the bulbs, the next step is to connect them all to either the Amazon Dot or the Amazon Echo. This will then be used with the other kinds of GE Link bulbs if you choose to use them. With this network connection for the Bluetooth to the lights, it can also be used for other objects in the house. It can also close blinds, open blinds, open doors, close doors, and switch lights on and off. All of this can be done through the Wink app.

To properly group the lights in your house after purchasing the needed resources, start by opening the Dot app and clicking on "Settings." Find "Connected Home." Add Wink to the list (it should be found in connected devices). Now you will be able to add more groups to the list. Once you have finished this process, you can try out the voice command for the house. For example, simply say "Alexa, turn on the hall lights." By doing this, you will be able to set the Alexa commands just as you did before. Also, you will be able to set a schedule for a program to turn off the lights or you can continue using the voice commands. By simply saying "Alexa, turn off all the lights," they can all be turned off once you get ready for bed.

To schedule the lights to go off at sunset or sunrise, you can set up a control by using a WeMo switch. These can be used without batteries and through the control of a Wi-Fi connection. You can control any type of light, such as fluorescent, halogen, LED, incandescent, and the fans with the WeMo switch. For this to be possible, start by downloading the WeMo app onto your own smartphone. Then connect the Wink Relay Channel to the WeMo switch channel on the IFTTT software. At this point, you must know that the relay channels will be able to fit well with the WeMo switches, the Android SMS, Tesco, Sensibo, Yo for Tesla, and EVE for Tesla. You will be able to control many items of work through the push of a button. After downloading the app and connecting the network, the rest becomes simple.

To turn the Hue lamps a different color, connect the Philips Hue channel to the IFTTT workflow channel. After this has been done, search through the options for creating your own kind of ambience with the use of a single command. Alexa will respond to your request through the device and will be able to turn the lights to different colors to create a soothing environment.

## Using Yonomi with Alexa

The next way to connect to the Amazon Dot or the Amazon Echo with Alexa is through the use of Yonomi. Alexa sometimes finds it difficult to apply herself to a task because of the amount of storage and the amount of schedules listed. Being able to switch routines on and off is not the best way for Alexa to work. Instead, there is a helper called Yonomi that works for scheduling and creating new forms of sheeting.

To set up Yonomi through Alexa, first open the Yonomi app and click on the icon that appears in the top left corner of the screen. Now, select the "Accounts and Hubs" tab. After that, tap on the bottom right-hand corner of the screen (or the upper right-hand corner if you are using iOS). From there, choose "Amazon Dot Account" and either log in or create an account with your own username and password. Now click on the "Connect" Button. The device will see the connection between the Yonomi and Alexa. To be sure of this, say the command "Alexa, discover my devices." After that, the app and everything else is set up and connected. Then you can continue to the next step of programming the routine.

To have Yonomi work for you, you have to write the programs for the software separately. The way to do this is by creating a normal daily routine that you go through. Then connect it to Yonomi and then you will be on your way. For example, pair the name of this with the room you want. Let's say, the hallway. Then you begin with the name Hall. Select the Sonos, the light bulbs, and the fan. Once you have done this, turn off the routine where you stop the Sonos and then turn off the light bulb and fan. Name this "Hall Off." The entire program started with the word "Hall" and ended with the phrase "Hall Off." If you go to the Dot discovery section of the app, it will show you both of these events and it will be able to connect the program to the device. After it is successfully processed through the system, you can use the voice command for the Hall lights. To turn them on, say "Alexa, turn on Hall." To turn them off, say "Alexa, turn off Hall." That is all it takes to get through the system and run the program. It may seem difficult at first, but at the end it becomes simple.

## WeMo Devices

There are a few more devices that have not been mentioned that can directly connect through the application. The first is the crockpot. Through this software, you can tell it to either cook slowly or turn off. If you are running around in the kitchen, sometimes saying the phrase is faster than pushing the button. Besides, it is more about how an upgrade can change life rather than offer a better way to do things. People normally upgrade their phones or computers for the smallest amount of increase in power and efficiency.

As before, there is also a switch to work through to control the device. By plugging the device into the WeMo Switch, lights or lamps controlled by the switch can be turned on or off when arriving home. With any other type of machine or maker, this device can also turn the devices on or off. It can let the appliance run smoothly through the appropriate command. This is still using short phrases and other techniques. It allows a person to speak quickly to have an action happen rather than walk around to turn off all the switches or run around to make sure that everything has been completely turned off. It is just another way to simplify a life system and provide more levels of efficiency with the surrounding network connection in your own smart home.

**Take Advantage of Its Skills**

You may be surprised to know that the Amazon Dot and the Alexa app have more specialized skills than mentioned before. If you would like to order some food, or you would like to request a ride, you can do this easily with the use of your Amazon Dot. You may even want to know what time your favorite shows are going to be shown. The best part about Amazon Dot is that more and more skills are being added and updated that you can use while performing day-to-day activities. Because new features are constantly being added, Amazon Dot and Alexa will be your dream partners in the near future.

# Chapter 5

# Connecting to a Bluetooth Speaker

If you would like to maximize the use of the Amazon Dot, you can connect it to another Bluetooth speaker.

**Some Things to Remember Before You Start Connecting:**

- Make sure that you place your Bluetooth speaker and your Amazon Dot at least three feet apart. That will allow Alexa to hear your voice commands. If you place the device too close to the speakers, it might not recognize your commands.
- Some speakers are certified to be perfect for connecting to the Amazon Dot. Using these speakers can be highly beneficial.

- Connect the speaker to other Bluetooth devices. A smartphone can be a good choice, although it can also be connected to other gadgets.
- Make sure you turn on the Bluetooth and the volume.
- The Amazon Dot can only connect to one Bluetooth device at a time. Make sure that you disconnect it from other Bluetooth devices before connecting it to your speaker.

## Connecting Your Amazon Dot to Your Chosen Bluetooth Speaker

You can now safely connect your Amazon Dot to the chosen Bluetooth speaker. Connecting speakers will be an easy task if you follow these few careful steps:

- Turn on your Bluetooth speaker and make sure that your Amazon Dot is ready to pair up with it. Most Bluetooth speakers come with a user guide that will give you the information that you need.
- Open the Alexa app and choose "Settings."
- Select the device that you are going to pair and then select "Bluetooth." The Amazon Dot will list all of the devices that you can connect; just connect it to your desired Bluetooth speaker.

- Once your Bluetooth speaker is connected, Alexa will let you know that you can connect successfully.
- If you would like to disconnect the Bluetooth speaker from your Amazon Dot, Just say your wake word and say "Disconnect."

## Tips on Using the Amazon Dot with Bluetooth Speaker

There are some tips that you have to remember so that you can easily use the speaker with the Amazon Dot.

- It is a good idea to control the volume of the Bluetooth speaker by using your voice; you can do this if your Bluetooth speaker is paired correctly with your Amazon Dot.
- If the Bluetooth speaker has already been connected to the device before, Just say "Connect" and it will automatically connect again.
- You can manage your settings by going to "Settings" and checking the device name.

*Andrew Mckinnon*

# Chapter 6
# Troubleshooting

You do know that there will be moments when the Amazon Dot will not work as well as you would like. You may encounter some issues but, before you panic, remember that these may be things that you can fix on your own. Check out some of the problems that may arise and how you can fix them.

## Echo Dot Doesn't Want to Connect to Other Speakers

The Amazon Dot can be paired with other speakers with the use of the 3.5mm audio cable or through Bluetooth but, in case it does not want to connect, here are some things that can be done:

- Try to connect the speaker to another device.

- If you are using battery-operated Bluetooth speakers, check if the battery is working properly.
- You may need to move the speaker away from the Amazon Dot to avoid interference.
- Using the Alexa app, forget the other speakers that you have connected to the Amazon Dot before. Once again, the Amazon Dot can only connect to one Bluetooth device at a time.
- Try reconnecting the Alexa app with the Bluetooth speaker.

If you are not going to use a Bluetooth speaker and you seek to connect the Amazon Dot by using the cable, but it does not work, here are some of the things that you can do:

- Check if the speaker that you are connecting it to is functioning properly.
- Check to see if there is anything that may be blocking the speaker.
- Try using an adapter.

**Alexa App Does Not Understand Your Command**

You have to remember that you cannot make Alexa work if you are not connected to the Internet. You always need to be connected to Wi-Fi so that you can voice your commands.

- Make sure that you the Amazon Dot is in an excellent location. It has to be somewhere where the device can hear your commands.
- It is not advisable to place your Amazon Dot on the floor; you need to put it somewhere up higher than your knee level.
- Be clear about your commands.
- Try to reduce background noises when you are talking to Alexa voice command.
- Repeat your question if it was not understood.
- Make your question or command more specific.

## The Amazon Dot Does Not Want to Connect to Wi-Fi

The Wi-Fi network has to be robust enough to be recognized by the device so that it will be connected correctly.

- Get to know the current status of your Wi-Fi network.

- Check if the speaker can detect the Wi-Fi network. You should be able to tell about this by looking at the color of the lights.

If your Amazon Dot is not connected to your Wi-Fi network, here are some things that you can do:

- Try to reconnect to the Wi-Fi network.
- Type the right password for your network.
- Check if your other devices can connect to the Wi-Fi network.
- You may want to disconnect other devices that are attached to your Wi-Fi network.
- Place your device closer to the router for a stronger Wi-Fi connection.

In case the things that are mentioned above do not work, here are other options:

- Manually set the Amazon Dot to connect to the Wi-Fi network by pressing the action button.
- Turn off your modem and restart it.
- Use the Alexa app to search for a Wi-Fi network.
- Try to connect the Amazon Dot to the Wi-Fi network again.

- If all these things do not work, it is best to contact the network provider.

*Andrew Mckinnon*

# Conclusion

Thank you again for downloading this book!

In short, Amazon Dot is a much smaller version of the Amazon Echo but it is also a great addition to Amazon's family of smart gadgets. Amazon Dot still uses the intelligence of the Echo, but in a more compact and portable device. Amazon Dot is a brilliant virtual assistant for those looking for a device that can not only play music but also give you today's forecast or read your favorite audio book. An even better feature of the Amazon Dot is that it is about half the price of its immediate sibling, the Amazon Echo, which makes it a more affordable choice. This Alexa-enabled device adapts to meet to your requirements and will be a great companion to other smart gadgets in your home.

I hope that you have understood everything that was written in the book and can begin to maximize the use of your new Amazon Dot or help you make an informed decision on whether to purchase it. The information that you have learned here is meant to guide you towards effective usage and enjoyment of your Amazing Amazon Dot for the first time!

Amazon Dot is an ideal choice for those who want to have a static speaker powered by a virtual assistant. Amazon Dot will fulfill your music needs for casual listening, or you can attach it to a more powerful external speaker when you want to "get loud."

If you have enjoyed the things that you have read this book, may I ask a favor from you?

Will you be kind enough to leave a review of this book on Amazon? I will be delighted to hear what you have to say!